Preface

We are pleased to present the "Science Olympiad All Chapters Combined MCQ Class 4", an engaging and comprehensive resource designed to help students excel in Science Olympiad examinations. This book offers a wide range of Multiple-Choice Questions (MCQs) covering all chapters of the Class 4 science syllabus, ensuring thorough preparation and conceptual understanding.

Science ignites curiosity and fosters problem-solving skills, and this book aims to make the learning process enjoyable and rewarding. Each question is carefully curated to be age-appropriate, challenging, and aligned with the Olympiad exam format, equipping students with the skills they need to succeed.

We hope this book serves as a valuable guide for Class 4 students, helping them excel in Olympiad competitions and fostering a lifelong love for science and discovery.

Table of Contents

Questions: Part 1

1. What is the primary function of the human skeleton?

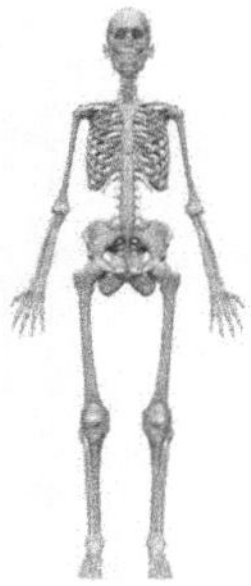

a) Digestion

b) Support

c) Respiration

d) Circulation

2. Which organ in the human body is primarily responsible for filtering blood?

a) Heart

b) Kidney

c) Liver

d) Lungs

3. Which planet is the smallest in our solar system?

a) Earth

b) Mercury

c) Mars

d) Venus

4. What type of clouds are white, fluffy, and often mean good weather?

a) Cumulus

b) Stratus

c) Cirrus

d) Nimbus

5. What part of the plant conducts photosynthesis?

a) Stem

b) Roots

c) Leaves

d) Flowers

6. What is the name of the process where liquid water turns into water vapor?

a) Precipitation

b) Evaporation

c) Condensation

d) Transpiration

7. What do we call the baby of a cow?

a) Kid

b) Calf

c) Lamb

d) Fawn

8. Which gas do we exhale?

a) Oxygen

b) Hydrogen

c) Nitrogen

d) Carbon dioxide

9. What is the main source of energy for plants?

a) Soil

b) Water

c) Sunlight

d) Air

10. What is the chemical formula for water?

a) H2O

b) CO2

c) O2

d) N2

11. What type of animal is a salamander?

a) Mammal

b) Reptile

c) Amphibian

d) Bird

12. What do you call the process by which plants lose water through small openings in their leaves?

a) Absorption

b) Respiration

c) Transpiration

d) Photosynthesis

13. Which planet is known for having a prominent ring system?

a) Jupiter

b) Uranus

c) Saturn

d) Neptune

14. What is the main component of the Earth's atmosphere?

a) Oxygen

b) Nitrogen

c) Carbon dioxide

d) Argon

15. What do we call the group of bones that protect the heart and lungs?

a) Skull

b) Spine

c) Rib cage

d) Pelvis

16. What is the largest organ inside the human body?

a) Brain

b) Liver

c) Heart

d) Kidney

17. Which part of a flower develops into fruit?

a) Petal

b) Stamen

c) Pistil

d) Sepal

18. What do we call animals that are active at night?

a) Diurnal

b) Nocturnal

c) Herbivores

d) Carnivores

19. What is the process of a liquid changing into a solid called?

a) Melting

b) Evaporation

c) Freezing

d) Condensation

20. Which of these animals is a herbivore?

a) Lion

b) Tiger

c) Elephant

d) Wolf

21. What is the smallest planet in our solar system?

a) Venus

b) Mercury

c) Earth

d) Mars

22. What do we call the part of the Earth where life exists?

a) Atmosphere

b) Lithosphere

c) Hydrosphere

d) Biosphere

23. What is the main purpose of a flower?

a) To make food

b) To reproduce

c) To absorb water

d) To protect the plant

24. Which organ in the human body is primarily responsible for pumping blood?

a) Liver

b) Lungs

c) Heart

d) Stomach

25. What do you call a scientist who studies rocks?

a) Chemist

b) Botanist

c) Zoologist

d) Geologist

26. Which planet is known as the Earth's twin?

a) Mars

b) Venus

c) Mercury

d) Jupiter

27. What do we call the movement of the Earth around its own axis?

a) Revolution

b) Orbit

c) Rotation

d) Axis

28. What do plants release into the air as a byproduct of photosynthesis?

a) Carbon dioxide

b) Nitrogen

c) Oxygen

d) Hydrogen

29. What is the main source of energy for the Earth?

a) The Moon

b) The Sun

c) The Stars

d) The Wind

30. What type of rock is formed by heat and pressure?

a) Igneous

b) Sedimentary

c) Metamorphic

d) Fossil

31. What do we call a baby horse?

a) Calf

b) Kid

c) Foal

d) Lamb

32. What is the term for animals that eat only plants?

a) Carnivores

b) Herbivores

c) Omnivores

d) Insectivores

33. What is the primary gas found in the Earth's atmosphere?

a) Oxygen

b) Nitrogen

c) Carbon dioxide

d) Hydrogen

34. Which planet is known for its Great Red Spot?

a) Earth

b) Jupiter

c) Mars

d) Saturn

35. What is the process by which plants make their food using sunlight?

a) Respiration

b) Transpiration

c) Photosynthesis

d) Digestion

36. What is the term for the bones in the human hand?

a) Carpals

b) Tarsals

c) Phalanges

d) Metacarpals

37. What is the natural satellite of the Earth?

a) Sun

b) Mars

c) Moon

d) Venus

38. Which planet is known as the Red Planet?

a) Earth

b) Mars

c) Jupiter

d) Saturn

39. What is the chemical symbol for gold?

a) Ag

b) Au

c) Fe

d) Pb

40. What type of animal is a kangaroo?

a) Reptile

b) Mammal

c) Amphibian

d) Bird

41. Which organ is responsible for removing waste from the blood?

a) Heart

b) Lungs

c) Kidneys

d) Liver

42. What is the process by which a caterpillar changes into a butterfly?

a) Metamorphosis

b) Photosynthesis

c) Transpiration

d) Respiration

43. What is the largest mammal in the world?

a) Elephant

b) Giraffe

c) Blue whale

d) Rhino

44. What do we call the study of the weather?

a) Biology

b) Geology

c) Meteorology

d) Zoology

45. What type of energy is produced by moving water?

a) Solar energy

b) Wind energy

c) Hydroelectric energy

d) Geothermal energy

46. What do you call a baby sheep?

a) Calf

b) Foal

c) Lamb

d) Kid

47. What is the boiling point of water in degrees Celsius?

a) 0°C

b) 50°C

c) 100°C

d) 150°C

48. Which planet is closest to the Sun?

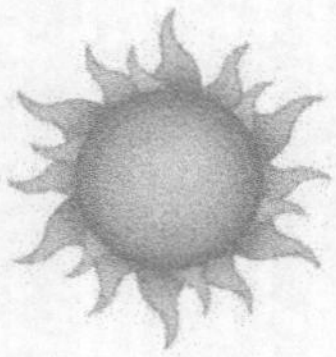

a) Venus

b) Earth

c) Mercury

d) Mars

49. What do we call the part of the plant that contains seeds?

a) Stem

b) Root

c) Flower

d) Leaf

50. What is the primary function of red blood cells?

a) Fight infection

b) Carry oxygen

c) Clot blood

d) Digest food

Questions: Part 2

51. What is the smallest planet in our solar system?

a) Venus

b) Mercury

c) Earth

d) Mars

52. What do we call the layer of the Earth where plants grow?

a) Crust

b) Mantle

c) Core

d) Lithosphere

53. What type of animal is a turtle?

a) Mammal

b) Reptile

c) Amphibian

d) Bird

54. What is the chemical symbol for oxygen?

a) O

b) Ox

c) O2

d) O3

55. What is the name of the galaxy we live in?

a) Andromeda

b) Milky Way

c) Triangulum

d) Whirlpool

56. What is the main function of white blood cells?

a) Carry oxygen

b) Fight infection

c) Clot blood

d) Provide nutrients

57. What is the term for the outermost layer of the Earth?

a) Crust

b) Mantle

c) Core

d) Lithosphere

58. Which part of the plant anchors it to the ground?

a) Leaf

b) Stem

c) Root

d) Flower

59. What type of rock is formed from magma or lava?

a) Sedimentary

b) Metamorphic

c) Igneous

d) Fossil

60. What is the term for the change of state from gas to liquid?

a) Evaporation

b) Condensation

c) Freezing

d) Sublimation

61. What do we call the part of a flower that attracts pollinators?

a) Stem

b) Leaf

c) Petal

d) Root

62. What is the name of the largest bone in the human body?

a) Skull

b) Rib

c) Femur

d) Spine

63. What is the main component of the human brain?

a) Bone

b) Muscle

c) Neurons

d) Blood

64. Which planet is known for having the most extensive ring system?

a) Jupiter

b) Uranus

c) Saturn

d) Neptune

65. What is the term for a scientist who studies weather patterns?

a) Geologist

b) Biologist

c) Meteorologist

d) Zoologist

66. What is the primary gas plants use during photosynthesis?

a) Oxygen

b) Nitrogen

c) Carbon dioxide

d) Hydrogen

67. What is the process by which plants lose water through their leaves?

a) Transpiration

b) Photosynthesis

c) Respiration

d) Absorption

68. What type of animal is a penguin?

a) Mammal

b) Bird

c) Fish

d) Reptile

69. What is the chemical symbol for iron?

a) Ir

b) Fe

c) Au

d) Ag

70. What is the natural force that causes objects to fall to the ground?

a) Magnetism

b) Friction

c) Gravity

d) Inertia

71. What is the term for a baby kangaroo?

a) Calf

b) Joey

c) Kid

d) Foal

72. What is the main function of the respiratory system?

a) Digestion

b) Respiration

c) Circulation

d) Excretion

73. Which planet is the hottest in our solar system?

a) Venus

b) Mercury

c) Mars

d) Jupiter

74. What type of energy is stored in a battery?

a) Solar energy

b) Electrical energy

c) Chemical energy

d) Kinetic energy

75. What is the process by which plants convert sunlight into food?

a) Transpiration

b) Respiration

c) Photosynthesis

d) Germination

76. What type of rock is limestone?

a) Igneous

b) Sedimentary

c) Metamorphic

d) Fossil

77. What do you call the young one of a cat?

a) Puppy

b) Kitten

c) Cub

d) Calf

78. What is the largest internal organ in the human body?

a) Heart

b) Lungs

c) Liver

d) Stomach

79. What is the term for animals that eat both plants and meat?

a) Herbivores

b) Carnivores

c) Omnivores

d) Insectivores

80. Which planet is known for its large red storm?

a) Mars

b) Jupiter

c) Saturn

d) Neptune

81. What is the term for the smallest unit of an element?

a) Atom

b) Molecule

c) Cell

d) Organ

82. What is the term for the bones that make up the fingers and toes?

a) Carpals

b) Metatarsals

c) Phalanges

d) Tarsals

83. What is the name of the largest ocean on Earth?

a) Atlantic Ocean

b) Indian Ocean

c) Arctic Ocean

d) Pacific Ocean

84. What is the main purpose of the roots in a plant?

a) Make food

b) Absorb water and nutrients

c) Support the plant

d) Reproduce

85. What is the term for a scientist who studies living organisms?

a) Geologist

b) Biologist

c) Physicist

d) Chemist

86. Which part of the plant contains chlorophyll?

a) Roots

b) Stem

c) Leaves

d) Flower

87. What type of animal is a frog?

a) Mammal

b) Reptile

c) Amphibian

d) Bird

88. What is the term for the force that resists the motion of objects?

a) Gravity

b) Magnetism

c) Friction

d) Inertia

89. What is the process of breaking down food in the body called?

a) Respiration

b) Digestion

c) Circulation

d) Excretion

90. What is the primary source of energy for the water cycle?

a) Wind

b) Sun

c) Moon

d) Earth's core

91. Which part of the flower produces pollen?

a) Petal

b) Pistil

c) Stamen

d) Sepal

92. What is the main function of the human heart?

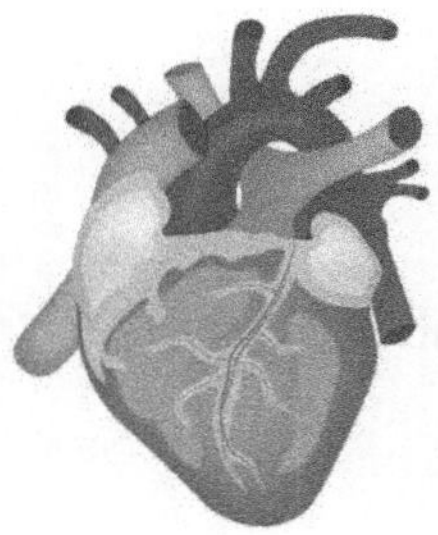

a) Digest food

b) Pump blood

c) Filter waste

d) Control movement

93. What is the term for a group of stars that form a pattern in the sky?

a) Galaxy

b) Constellation

c) Solar system

d) Universe

94. What is the chemical formula for carbon dioxide?

a) CO

b) CO2

c) C2O

d) CO3

95. What is the term for an animal that primarily eats insects?

a) Herbivore

b) Carnivore

c) Omnivore

d) Insectivore

96. Which planet has the most extensive ring system?

a) Jupiter

b) Saturn

c) Uranus

d) Neptune

97. What is the process by which an organism makes more of its kind?

a) Respiration

b) Reproduction

c) Digestion

d) Circulation

98. What is the term for water falling from the sky in the form of rain, snow, sleet, or hail?

a) Evaporation

b) Condensation

c) Precipitation

d) Sublimation

99. What type of energy is produced by the Sun?

a) Wind energy

b) Solar energy

c) Chemical energy

d) Nuclear energy

100. What is the largest planet in our solar system?

a) Earth

b) Venus

c) Jupiter

d) Saturn

Questions – Part 3

101. Which gas do humans breathe in for survival?

a) Carbon dioxide

b) Oxygen

c) Nitrogen

d) Hydrogen

102. Which planet is known as the Red Planet?

a) Venus

b) Jupiter

c) Mars

d) Mercury

103. Which part of the plant makes food?

a) Stem

b) Roots

c) Leaves

d) Flowers

104. What is the process of water changing into vapor called?

a) Condensation

b) Evaporation

c) Precipitation

d) Filtration

105. Which is the largest organ in the human body?

a) Heart

b) Skin

c) Brain

d) Liver

106. What do bees collect from flowers to make honey?

a) Pollen

b) Nectar

c) Water

d) Seeds

107. Which force pulls objects toward the Earth?

a) Friction

b) Gravity

c) Magnetism

d) Electricity

108. Which is the main source of energy for Earth?

a) Moon

b) Sun

c) Stars

d) Wind

109. What do we call animals that eat both plants and animals?

a) Herbivores

b) Carnivores

c) Omnivores

d) Decomposers

110. Which is the fastest land animal?

a) Cheetah

b) Lion

c) Tiger

d) Horse

111. Which planet is closest to the Sun?

a) Earth

b) Mercury

c) Mars

d) Venus

112. Which blood cells help fight infections?

a) Red blood cells

b) White blood cells

c) Platelets

d) Plasma

113. Which type of star is the Sun?

a) Red giant

b) Yellow dwarf

c) Neutron star

d) Supernova

114. Which part of the human body helps in breathing?

a) Stomach

b) Lungs

c) Liver

d) Kidneys

115. What is the boiling point of water?

a) 50°C

b) 80°C

c) 100°C

d) 120°C

116. Which is the hardest natural substance on Earth?

a) Gold

b) Diamond

c) Iron

d) Silver

117. What is the study of stars and planets called?

a) Biology

b) Astronomy

c) Geology

d) Chemistry

118. Which gas do plants release during photosynthesis?

a) Carbon dioxide

b) Oxygen

c) Nitrogen

d) Hydrogen

119. Which part of the Earth is made up of molten rock?

a) Crust

b) Mantle

c) Core

d) Surface

120. Which organ pumps blood through the body?

a) Brain

b) Heart

c) Lungs

d) Liver

Answers

1. b) Support

2. b) Kidney

3. b) Mercury

4. a) Cumulus

5. c) Leaves

6. b) Evaporation

7. b) Calf

8. d) Carbon dioxide

9. c) Sunlight

10. a) H2O

11. c) Amphibian

12. c) Transpiration

13. c) Saturn

14. b) Nitrogen

15. c) Rib cage

16. b) Liver

17. c) Pistil

18. b) Nocturnal

19. c) Freezing

20. c) Elephant

21. b) Mercury

22. d) Biosphere

23. b) To reproduce

24. c) Heart

25. d) Geologist

26. b) Venus

27. c) Rotation

28. c) Oxygen

29. b) The Sun

30. c) Metamorphic

31. c) Foal

32. b) Herbivores

33. b) Nitrogen

34. b) Jupiter

35. c) Photosynthesis

36. a) Carpals

37. c) Moon

38. b) Mars

39. b) Au

40. b) Mammal

41. c) Kidneys

42. a) Metamorphosis

43. c) Blue whale

44. c) Meteorology

45. c) Hydroelectric energy

46. c) Lamb

47. c) 100°C

48. c) Mercury

49. c) Flower

50. b) Carry oxygen

51. b) Mercury

52. d) Lithosphere

53. b) Reptile

54. a) O

55. b) Milky Way

56. b) Fight infection

57. a) Crust

58. c) Root

59. c) Igneous

60. b) Condensation

61. c) Petal

62. c) Femur

63. c) Neurons

64. c) Saturn

65. c) Meteorologist

66. c) Carbon dioxide

67. a) Transpiration

68. b) Bird

69. b) Fe

70. c) Gravity

71. b) Joey

72. b) Respiration

73. a) Venus

74. c) Chemical energy

75. c) Photosynthesis

76. b) Sedimentary

77. b) Kitten

78. c) Liver

79. c) Omnivores

80. b) Jupiter

81. a) Atom

82. c) Phalanges

83. d) Pacific Ocean

84. b) Absorb water and nutrients

85. b) Biologist

86. c) Leaves

87. c) Amphibian

88. c) Friction

89. b) Digestion

90. b) Sun

91. c) Stamen

92. b) Pump blood

93. b) Constellation

94. b) CO2

95. d) Insectivore

96. b) Saturn

97. b) Reproduction

98. c) Precipitation

99. b) Solar energy

100. c) Jupiter

101. b) Oxygen

102. c) Mars

103. c) Leaves

104. b) Evaporation

105. b) Skin

106. b) Nectar

107. b) Gravity

108. b) Sun

109. c) Omnivores

110. a) Cheetah

111. b) Mercury

112. b) White blood cells

113. b) Yellow dwarf

114. b) Lungs

115. c) 100°C

116. b) Diamond

117. b) Astronomy

118. b) Oxygen

119. b) Mantle

120. b) Heart

9 789369 72445

Printed at Repro India Ltd.